178477 EN
Crawling Centipedes

Bishop, Celeste
ATOS BL 1.7
Points: 0.5

LG

MW00846359

ICKY ANIMALS!
Small and Gross

CRAWLING CENTIPEDES

Celeste Bishop

PowerKiDS press™

New York

Published in 2016 by The Rosen Publishing Group, Inc.
29 East 21st Street, New York, NY 10010

First Edition

Editor: Sarah Machajewski
Book Design: Mickey Harmon

Photo Credits: Cover (centipede) Kittikorn Phongok/Shutterstock.com; cover, pp. 1, 3, 4, 7 ,8, 11, 12, 15, 16, 19, 20, 23, 24 (splatters) GreenBelka/Shutterstock.com; p. 5 Ryan M. Bolton/Shutterstock.com; p. 6 Nikolay Bassov/Shutterstock.com; p. 9 Lippert Photography/Shutterstock.com; p. 10 kalacs/Shutterstock.com; pp. 13, 24 (feelers) reptiles4all/Shutterstock.com; p. 14 c photospirit/Shutterstock.com; p. 17 Suede Chen/Shutterstock.com; p. 18 Anton Chernenko/Shutterstock.com; p. 21 BOONCHUAY PROMJIAM/Shutterstock.com; p. 22 Lewis Tse Pui Lung/Shutterstock.com; p. 24 (spider) DenRz/Shutterstock.com.

Library of Congress Publication Data

Bishop, Celeste, author.
  Crawling centipedes / Celeste Bishop.
      pages cm. — (Icky animals! Small and gross)
  Includes index.
  ISBN 978-1-4994-0692-4 (pbk.)
  ISBN 978-1-4994-0694-8 (6 pack)
  ISBN 978-1-4994-0696-2 (library binding)
  1. Centipedes—Juvenile literature. I. Title. II. Series: Bishop, Celeste. Icky animals! Small and gross.
  QL449.5.B57 2015
  595.6'2—dc23
                              2014048525

Manufactured in the United States of America

CPSIA Compliance Information: Batch #WS15PK: For Further Information contact Rosen Publishing, New York, New York at 1-800-237-9932

CONTENTS

Have you ever seen a centipede? Centipedes are small and creepy!

5

Centipedes are a kind of bug. Their body is long and flat. Most are red or brown.

The name "centipede" comes from words that mean "hundred" and "foot."

9

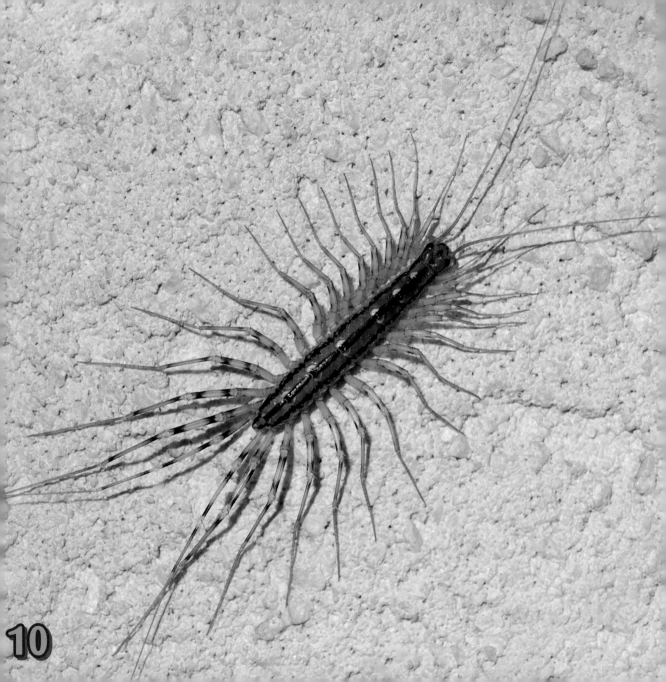

Some kinds of centipedes have over 300 legs!

Centipedes have **feelers** on their head. They're very long!

feelers

Feelers help centipedes sense what's around them.

Centipedes can't see well.
They use other senses
to find their way.

17

Centipedes eat **spiders** and other bugs. They think bugs are yummy!

Birds like to eat centipedes. Centipedes hide under rocks and leaves to stay safe.

21

If a centipede sees you coming, it might run away! Its many legs help it move fast.

WORDS TO KNOW

feelers

spider

INDEX

WEBSITES

Due to the changing nature of Internet links, PowerKids Press has developed an online list of websites related to the subject of this book. This site is updated regularly. Please use this link to access the list: www.powerkidslinks.com/icky/cent